Rafael Nascimento de Assis
Márcia Brazão

Molecular identification of the CTX-M resistance gene in bacteria

Rafael Nascimento de Assis
Márcia Brazão

Molecular identification of the CTX-M resistance gene in bacteria

Bacterial resistance

ScienciaScripts

Cover image: www.ingimage.com

This book is a translation from the original published under ISBN 978-613-9-65078-1.

Publisher:
Sciencia Scripts
is a trademark of
Dodo Books Indian Ocean Ltd. and OmniScriptum S.R.L publishing group

120 High Road, East Finchley, London, N2 9ED, United Kingdom
Str. Armeneasca 28/1, office 1, Chisinau MD-2012, Republic of Moldova, Europe
Printed at: see last page
ISBN: 978-620-7-84973-4

I dedicate this work first of all to God, for his strength and divine protection. Especially to my mother Aurilene Lopes do Nascimento and my father François Dutra de Assis, for their total support throughout this journey.

ACKNOWLEDGEMENTS

I would firstly like to thank God for giving me the strength to always persist in my goals. To my entire family for all their support and for always believing in my potential. To the friends who have been by my side since the beginning of the course, Wanderley José da Silva Júnior, Yasmim Picanço Granjeiro, Ramila Santana Araújo, Ana Karoline, Djádila Rayane Oliveira and my dear Glaciane Rocha da Silva who was, without a doubt, the one who was with me the most and helped me at all times with her words of encouragement, her advice and her tugs on the ear.To Enayra Christina Clemente Ferreira for her great interest in helping everyone, for being the person most responsible for always getting everyone together and maintaining friendships, for all her help in carrying out this research and, above all, for her friendship.

To Ismael Alexandre da Silva Nascimento for his excellent guidance, conversations and moments of relaxation. To my dear research sister Edimilla Carneiro da Cunha Aguiar for her affection, advice, help in the development of the research, conversations and laughter.To the entire LaBMol team, Juliana Aline de Souza Lemos, Claudiane Raquel, Kemeson Alves Naveca, Miguel Ferreira Dias, Wilson, Matheus Victor de Almeida Xavier, Andréa Souza dos Santos, Aline Felippin Lima, Débora Dinelly and Raíssa, for having contributed to my knowledge throughout this journey.I undoubtedly have a lot to thank each and every one of these people for, above all, giving me the pleasure of enjoying great moments. We really enjoyed the best that the University had to offer us, and when I think about what that best would be, I can't think of anything other than the FRIENDSHIP of all of you.

To my dear teacher and supervisor Fabiana Granja for always believing in my potential, giving me challenges, making me overcome my difficulties and for all the knowledge she passed on.To my dear co-supervisor Márcia Brazão e Silva Brandão, who together with her husband Derlano Bentes Capucho infected me with all their enthusiasm and sparked my interest in research, for all their support and trust.To all my teachers for contributing to my learning in the Universe of Biological Sciences.

SUMMARY

The CTX-M gene originated as an intrinsic resistance to beta-lactams present in the chromosomes of bacteria of the genus *Kluyvera spp.* that codes for an extended-spectrum beta-lactamase enzyme (ESBL). It gives the bacteria that produce it the ability to render broad-spectrum beta-lactam antimicrobials useless. Its hydrolytic action can be interrupted by the use of associated inhibitors such as clavulanic acid, sulbactam and tazobactam. The aim of this study was to identify the CTX-M resistance gene in ESBL-positive bacteria from the state of Roraima using molecular tests. The bacterial samples used came from various biological materials such as catheter tips, wound secretions, sputum, blood and urine from hospitals in the state of Roraima. From the clinical material, the strains were seeded and isolated on MacConkey agar for the phenotypic identification of Gram-negative strains using biochemical tests. The disc diffusion test was carried out to verify the presence of ESBL. For molecular identification of the CTX-M gene, the DNA of ESBL-positive strains was extracted by heat-shocking the samples and exposing them to boiling. Polymerase Chain Reaction (PCR) was then carried out on the samples obtained using specific primers that delimited the region of plasmid DNA of approximately 593pb that encodes the CTX-M enzyme and confirmed by agarose gel electrophoresis. *Klebsiella pneumoniae* (17CTXLaBMol/UFRR) was used as a positive control. Among the 72 strains that tested positive for ESBL in the disc diffusion test, we found the molecular presence of the CTX-M resistance gene in 29% of the samples. We observed similar results for the presence of the gene among the different biological materials. Among the bacteria that presented the CTX-M gene, we observed a predominance of the Klebsiella genus with 61.9% representatives. When analysing the resistance/susceptibility profile of the strains to antimicrobials, we observed greater susceptibility to amikacin, imipenem, levofloxacin, piperacillin + tazobactam and chloramphenicol.

Keywords: Bacteria. Resistance. ESBL. Gram-negative. Antimicrobials.

SUMMARY

CHAPTER 1

INTRODUCTION

1.1 PRESENTATION

The CTX-M resistance gene produces Extended Spectrum Beta-lactamase (ESBL) enzymes that are capable of inhibiting the action of broad-spectrum beta-lactam antimicrobials by irreversibly hydroxylating the amide bond of the beta-lactam ring of these antimicrobials.

In general, the gene that codes for ESBLs is located on plasmids, and these enzymes are characterised by their ability to inactivate beta-lactam antimicrobials such as penicillins, cephalosporins and monobactams, with the exception of carbapenems. Their hydrolytic action on antimicrobials can be interrupted with the use of beta-lactamase inhibitors such as clavulanic acid, sulbactam and tazobactam.

The CTX-M gene originated from the mobilisation of beta-lactam resistance genes present in the chromosomes of *Kluyvera* spp. bacteria, which from point mutations derived into different subtypes of the gene, in response to the selective pressure intensified by the widespread indiscriminate use of beta-lactams.

The increase in the number of patients hospitalised in the state's public health system who do not respond to infectious treatments suggests that the presence of antimicrobial-resistant strains is becoming increasingly frequent, which is causing a serious public health problem.

Therefore, there is a need to carry out antimicrobial resistance/susceptibility tests on ESBL enzyme-producing bacterial species related to infections, isolated from samples from all over the state of Roraima, and among these, the identification of the CTX-M resistance gene.

Research involving resistant strains in Roraima's public hospitals will contribute to medical practice by indicating which bacterial species are more susceptible to the presence of the gene and to treatment with more appropriate antimicrobials, enabling patients to improve quickly, reducing hospital stays and treatment costs, as well as helping to minimise the effects of the selection of resistant micro-organisms.

In view of the above, the aim of this study was to determine the frequency of the CTX-M gene in Gram-negative bacteria from July 2015 to July 2016, after identifying the species that were ESBL-positive.

1.2 LITERATURE REVIEW

1.2.1 Gram-negative bacteria

The group of prokaryotes includes bacteria and archaeobacteria. These organisms are unicellular and are mainly characterised by the absence of a karyotheca, organelles and the presence of a cell wall. However, from a metabolic point of view, these organisms are complex and diverse, which allows them to adapt to varied environments (TORTORA, FUNKE, CASE, 2010).

Among the prokaryotes, bacteria are the main representatives, classified into two large groups according to the structure of their cell wall. Gram-positive bacteria have a thick layer of peptidoglycan, which is a polysaccharide polymer made up of alternating layers of N-acetyl-glucosamine and N-acetyl-muramic acid, interconnected by tetrapeptides (ARAÚJO, 2010; BEVERIDGE, 1999).

Gram-negative bacteria have a thin layer of peptidoglycan that is also associated with an outer membrane made up of proteins, phospholipids and lipopolysaccharide (LPS), the latter of which is responsible for provoking a strong response from the host's immune system and is the main factor that characterises the group's high degree of pathogenicity and associates it with hospital outbreaks (ARAÚJO, 2010; BEVERIDGE, 1999; CRUZ-MACHADO, 2010). Figure 1 shows the difference in the wall structure of both groups of bacteria.

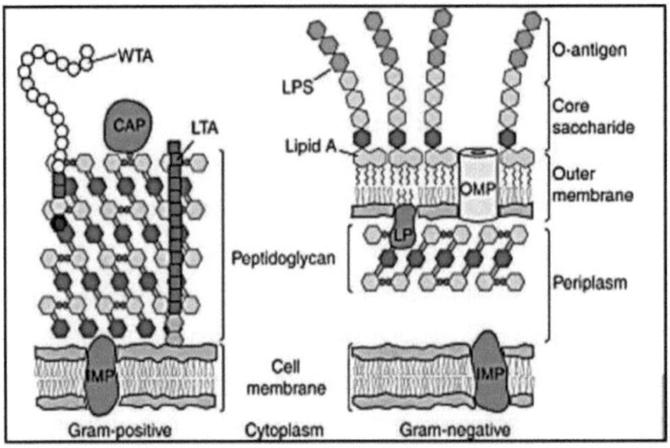

Figura 1- Cell wall structure in Gram-Positive and Negative bacteria

Source: Adapted from Silhavy et al., (2010).

Gram-negative bacteria are commonly related to resistance to various drugs, since they have an outer membrane that reduces the entry of these components and prevents them from reaching their target, an event that does not occur in Gram-positive bacteria (TORTORA, FUNKE, CASE, 2010).

Another worrying factor associated with Gram-negative infections is the scarcity of drugs for treatment and their great ability to acquire resistance to most existing drugs. In contrast, there is a range of drugs that are effective in treating Gram-positive bacteria and the absence of an outer membrane in this group may explain their susceptibility to these drugs (BUSH, 2010).

The emergence of resistance to antimicrobials is primarily related to undirected mutations which, through the process of natural selection, make organisms that are better able to survive the environment stand out. However, drugs indicated indiscriminately by doctors or through self-medication by the patient act as selective agents that accelerate the process of natural selection, creating a more favourable environment for the growth of resistant bacteria (ANTONIO, 2009).

In general, Gram-negative bacteria are resistant to most penicillins because they cannot overcome the barrier formed by the outer membrane and thus reach the peptidoglycan tetrapeptide cross-links and prevent the bacteria from growing (TORTORA, FUNKE, CASE, 2010).

These characteristics have linked Gram-negative bacteria to serious cases of infection, making them a worrying pathogen. The selective pressure caused by the intense use of antimicrobials used in infectious treatments has led to the emergence of mechanisms that inactivate different types of drugs in these microorganisms, making them the main group related to hospital outbreaks (CRUZ-MACHADO, 2010; KORZENIEWSKA, HARNISZ, 2013; UMADEVI et al., 2011).

1.2.2 Enterobacteriaceae

Among pathogenic Gram-negative bacteria, the Enterobacteriaceae family stands out as an important group of pathogens that inhabit the gastrointestinal tract of humans and other animals and can cause infections in their host, as well as representing one of the families with the greatest capacity to acquire resistance (FARINAS, MARTÍNEZ-MARTÍNEZ, 2013).

Among the Enterobacteriaceae, the genera Citrobacter, Enterobacter, Escherichia, Klebsiella, Proteus, Salmonella, Shigella and Serratia stand out as the main ones related to infections in the urinary tract, respiratory tract, blood and wounds, as well as outbreaks of multidrug resistance (FARINAS, MARTÍNEZ-MARTÍNEZ, 2013).

The *Klebsiella* genus is described by Lima et al. (2014) as the main group related to hospital outbreaks. They are non-motile Gram-negative bacilli that are usually encapsulated. The genus comprises five species, *K. planticola, K. terrigena and K. ornithinolytica, K. oxytoca, K. pneumoniae* and *K. ozaenae,* the latter being considered a subspecies of *K. pneumoniae* (MOREIRA, 2014).

The *Klebsiella pneumoniae* and *Escherichia coli* species stand out for being reported as the main strains carrying resistance genes involved in the production of enzymes capable of inhibiting various drugs, making them the main species involved in hospital outbreaks (FARINAS, MARTÍNEZ-MARTÍNEZ, 2013).

Among Gram-negative species not belonging to the Enterobactereceae family, the presence of resistance genes that code for antimicrobial inhibitory enzymes is less common. However, *P. aeruginosa* is one of the examples of pathogens that most cause hospital-acquired infections and is characterised by having high intrinsic resistance, as well as acquired resistance, generating multidrug resistance, which implies a high degree of pathogenicity (ALMEIDA et aL, 2012).

1.2.3 Structures that influence virulence

As for the virulence of Gram-negative bacteria, they have various structures that can facilitate infection of the host and thus make them more virulent (CRUZ-MACHADO, 2010).

The cell wall, for example, determines the shape of the cell and its main function is to protect it from rupture due to the effects of osmotic pressure, as well as acting as a barrier for various drugs. Its influence is related to the function of anchoring structures such as flagella, axial filaments and toxins (TORTORA, FUNKE, CASE, 2010).

Lipopolysaccharides (LPS) act as endotoxins for the host and are found on the outer membrane of bacteria, being the main factor responsible for the pathogenicity and high virulence of Gram-negative bacteria (CRUZ-MACHADO, 2010; MACHADO et aL, 2004).

These endotoxins, as shown in figure 2, are made up of lipid A, which will act as a toxin and will be responsible for triggering an immune response in the host when it is released with the death of the microorganism. The polysaccharide core will provide stability to the molecule and the polysaccharide O will act as an antigen, which will be recognised by the host's antibody (MACHADO et aL, 2004).

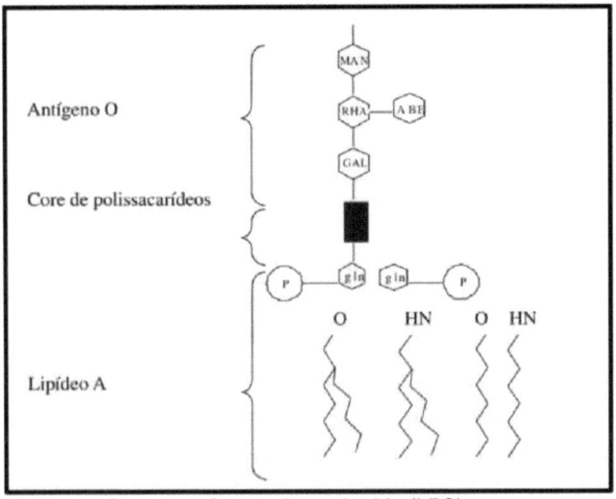

Figura 2- Structure of Lipopolysaccharide (LPS)

Source: Araújo (2010).

This structure gives the LPS a high antigenic capacity and causes it to provoke inflammatory reactions due to a strong immune system response, such as fever and dilation of venous vessels (CRUZ-MACHADO, 2010; MACHADO et al., 2004).

In more serious cases, as a result of generalised infections, the pathogen in the bloodstream causes septic shock due to the endotoxins that are released, which are related to high mortality rates (MACHADO et al, 2004).

The glycocalyx, present in some species of bacteria, is a viscous polymer external to the cell wall, made up of polysaccharide, polypeptide or both, which are related to protection against phagocytosis, dehydration and adherence to surfaces. This structure can be strongly adhered to the cell wall and is called a capsule, or it can be loosely adhered and is described as a viscous layer (TORTORA, FUNKE, CASE, 2010).

Flagella allow bacteria to move to favourable environments for growth. They are made up of three regions: the outermost filament is formed by a globular protein, flagellin; attached to this is a hook and the basal body responsible for anchoring the flagellum to the cell wall and plasma membrane (TORTORA, FUNKE, CASE, 2010).

The fimbriae, formed by the pilin protein, with a surface attachment function, allow the bacteria to colonise mucous surfaces. To a lesser extent, it is related to locomotion (TORTORA, FUNKE, CASE, 2010).

The sexual *pili*, specific to Gram-negative bacteria, also formed by the pilin protein, is the main structure that enables the dissemination of resistance through the process called conjugation. This dissemination influences the virulence of these organisms (TORTORA, FUNKE, CASE, 2010).

In addition, several other factors that influence the virulence of bacteria are still unknown. However, these structures allow the bacteria to overcome the host's defences and thus cause the disease and are related to their great capacity for adaptation (MARTINEZ, 2009).

1.2.4 Antimicrobials

Among the groups of antimicrobials used to treat infection by Gram-negative bacteria, the beta-lactam antimicrobials stand out, which include penicillins, cephalosporins (divided into 4 groups, called 1ª , 2ª , 3ª and 4ª generation); monobactams and carbapenems, as these are more effective at inhibiting bacterial growth (CARMO et al., 2012).

These drugs pass through the outer membrane and reach the site of bacterial growth inhibition, an action that characterises their high efficiency in controlling bacterial growth (CARMO et al., 2012). The figure below shows the classes of beta-lactams with their main antimicrobials.

Figura 3- Classes of beta-lactams and their main antimicrobials

Class	Antimicrobials
Penicillins	Ampicillin, amoxicillin, oxacillin and piperacillin.

Cephalosporins	**1° generation:** Cephalexin, cefazolin. **2° generation:** Cefoxitin, cefuroxime. **3° generation:** Cefotaxime, ceftazidime, ceftriaxone. **4° generation:** Cefepime, cefpiroma.
Monobactams	Aztreonam.
Carbapenems	Ertapenem, imipenem and meropenem.

Source: Adapted from Bush (2010).

Most beta-lactam antimicrobials have a bactericidal action, inhibiting bacterial growth by preventing the formation of cross-links between peptidoglycans, thus inhibiting cell wall synthesis (CARMO et al., 2012). They are the most widely used antimicrobials in infectious treatments due to their high efficiency and low toxicity (AHMED et al., 2013; ZARFEL et al., 2013). Their chemical structure is characterised by a beta-lactam ring, shown in figure 4, made up of a cyclic amide, in which the chain is closed by binding the nitrogen of the amide to the carbon in the beta position, giving rise to the beta-lactam name of the antimicrobial (AHMED et al., 2013).

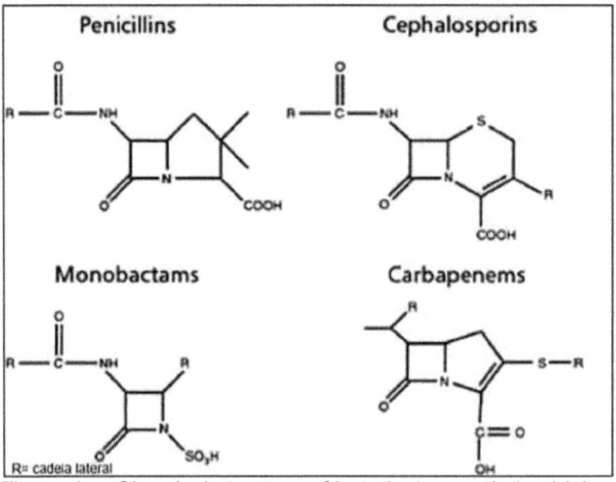

Figura 4- - Chemical structure of beta-lactam antimicrobials

Source: Williams (1999).

1.2.5 Emergence of Extended Spectrum Beta-Lactamase (ESBL) enzymes

Bacteria adopt various mechanisms that make them resistant to different drugs, such as altering membrane permeability and altering transport systems in the cell. In addition, there are mechanisms for removing the drug from the intracellular environment by so-called efflux pumps, altering the target site of the antimicrobial, modifying and synthesising metabolic pathways to escape the effects of antimicrobials and inactivating antimicrobials by synthesising enzymes (OLIVEIRA, SILVA, 2009; TORTORA, FUNKE, CASE, 2010).

The production of enzymes that inhibit the action of antimicrobials is the most important mechanism of resistance of Gram-negative bacteria, and is strongly related to hospital outbreaks, interfering with patient treatment due to its broad capacity to inhibit the drug (CARTAXO, 2015).

In 1980, beta-lactam antibiotics with a broad spectrum of action appeared in response to the widespread presence of beta-lactamase enzymes in
pathogens, which inhibited the action of the simplest beta-lactams (narrow spectrum) available. The selective pressure exerted by the intense use of broad-spectrum beta-lactams has contributed to the prominence of bacteria carrying *Extended Spectrum* Beta-Lactamase (ESBL) enzymes (AHMED et aL, 2013).

The effect of ESBLs in inhibiting most of the existing beta-lactam antimicrobials is also associated with the intrinsic pathogenicity of Gram-negative bacteria, especially Enterobacteriaceae, making them the predominant form of defence in this group (KORZENIEWSKA; HARNISZ, 2013; UMADEVI et aL, 2011).

1.2.6 Genotypic characterisation of ESBLs

ESBLs can be classified according to Ambler's molecular scheme (1980), which divides beta-lactamases into four groups (A, B, C, D), in which classes A, C and D are represented by enzymes that have a serine in the enzyme's active site as their main catalytic residue. Class (B) or Metallo- Beta-Lactamases are represented by enzymes that have zinc as their main catalytic residue. The vast majority of ESBLs fall into Ambler's molecular class (A) (DALMARCO, BLATT, CÓRDOVA, 2006; OZTURKL, OZKIRIMILI, OZGUR, 2015).

The first ESBL genotypes identified originated from genetic mutations of restricted-spectrum enzymes (TEM-1, TEM-2 and SHV-1) of the TEM (Temoniera) and SHV (Sulfhydryl Variable) types. These groups of enzymes evolved through changes in the amino acid sequence of the enzyme's active site, increasing its hydrolytic capacity (CARMO et aL, 2012; SILVA, SILVA, OLIVEIRA, 2014). However, new types of ESBLs have emerged in recent years, including the CTX-M enzyme, which has the largest number of variants described in recent years (FOURNIER et aL, 2013).

1.2.7 Inactivation of beta-lactam antimicrobials by ESBL enzymes

ESBLs are enzymes characterised by inhibiting beta- lactam antimicrobials, penicillins, first to fourth generation cephalosporins and monobactams, but they are not able to inhibit the action of carbapenems, as they have a broad spectrum of activity against Gram-negative strains (AHMED et al, 2013; ZARFEL et al, 2013). To date, there are more than 200 genes that code for different types of ESBLs (KORZENIEWSKA, HARNISZ, 2013; ZARFEL et aL, 2013). The inactivation mechanism occurs when the hydroxyl linked to the serine amino acid present in the active site of these enzymes binds irreversibly to the beta-lactam ring of the antimicrobial (figure 5), rendering it devoid of functional biological activity (AHMED et aL, 2013).

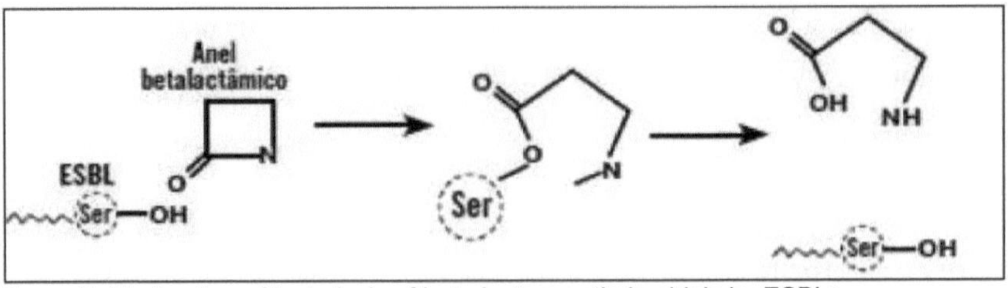

Figure 5- Mechanism of hydrolysis of beta-lactam antimicrobials by ESBLs
Source: Silva (2012).

However, this group of enzymes is inhibited by clavulanic acid, sulbactam and tazobactam, shown in figure 6 (AHMED et aL, 2013; MEYER, PICOLI, 2011). These inhibitors are similar in structure to penicillins and act as substrates for the active sites of ESBL enzymes, allowing the administered antimicrobial to remain available to perform its pharmacological function (NEVES et al., 2011).

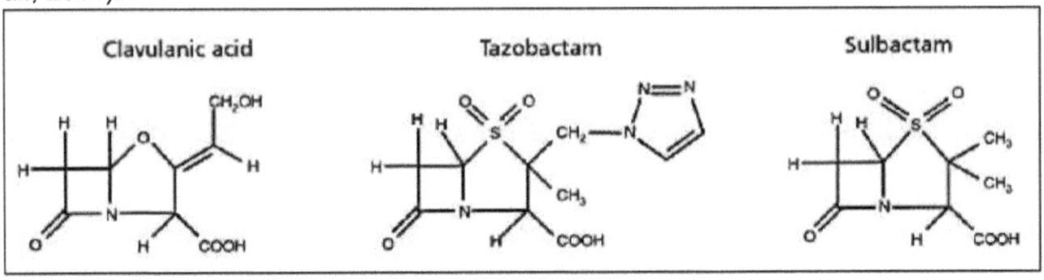

Figure 6- Extended spectrum beta-lactamase (ESBL) inhibitors
Source: Williams (1999).

ESBLs can be encoded by genes present in chromosomes and plasmids, and can also be present in transposons and integrons, which mediate mobilisation between plasmids and chromosomes (OLIVEIRA, SILVA, 2009).

Studies related to the most frequent bacterial type as ESBL producers indicate *Klebsiela pneumoniae* and *E. coli* (UMADEVI et aL, 2011), as well as *Proteus mirabilis, Enterobacter* sp., *Acinetobacter* sp., *Pseudomonas sp.* (SILVA, SILVA, OLIVEIRA, 2014; ZARFEL et aL, 2013).

Although most of the outbreaks of ESBL-producing strains have been reported in hospital environments, their presence is not restricted to these areas, but can also be present in the community, for example in sewage, domestic waste, animal faeces and food (UMADEVI et al., 2011; ZARFEL et al., 2013).

1.2.8 Genetic mechanism of resistance: Horizontal gene transfer

Bacteria have intrinsic resistance mechanisms derived from vertical gene transfer, i.e. from parent to offspring, and also by horizontal transfer mechanisms, such as forms of gene recombination, which refer to the transfer of genes to individuals of the same generation (DANTAS, SOMMER, 2012; MARTINEZ, 2009).

The transfer of resistance genes is commonly associated with mobile genetic elements such as plasmids, transposons, integrons, and/or insertion sequences (ISCRs), which allow both the mobilisation of the gene between

chromosomes and plasmids of the same strain, as well as between different bacteria and even in bacteria that are poorly related (WELLINGTON et al., 2013; BLAIR, 2015).

Plasmids are the main mobile genetic elements and are characterised by being extrachromosomal elements with genes that code for bacterial resistance. They are strongly associated with the dispersal of resistance between different bacterial groups, since they have autonomous replication and horizontal transfer capacity (BROLUND et al., 2013).

They are double-stranded circular DNA fragments that replicate independently of the bacterial cell and have characteristics that allow the organism to survive in different environments. However, this molecule is not essential for the life of these organisms (BROLUND et al., 2013).

Transposons, like plasmids, are considered to be mobile genetic elements. The cleavage and mobility of these fragments are conferred by genes that express transposase enzymes. Once cleaved, they can be translocated to another region of the molecule of origin or to another molecule. For example, from the chromosome to the plasmid (OLIVEIRA, SILVA, 2009; TORTORA, FUNKE, CASE, 2010).

Figure 7 illustrates the mechanism of transfer of resistance genes (in red) from a chromosome to the plasmid and from there, the spread to other bacteria.

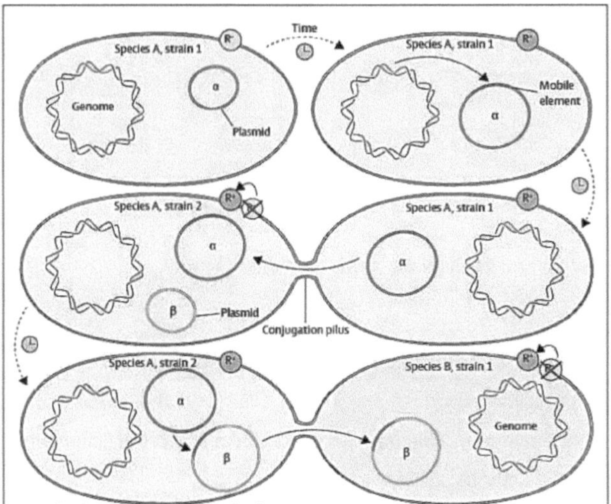

*R$^-$ = fenótipo sensível. R$^+$ = fenótipo resistente.

Figura 7- Mobilisation of the resistance gene

Source: Wellington et al., (2013).

Integrons consist of an *int* gene, which encodes a specific recombination enzyme called an integrase at a site containing short DNA sequences called gene cassettes and a binding site for these cassettes called an *att.* These are mobile elements that include a single gene (most commonly an antimicrobial resistance gene) and are inserted into the integron by the integrase enzyme (OCHMAN et al., 2000).

Because it has the ability to move from one integron to another or to different locations in the same integron, a gene cassette is characterised as a small, autonomous, non-replicating, circular double-stranded DNA molecule that lacks promoters to express the genes it carries (OCHMAN et al, 2000).

Insertion sequences are elements that carry genes capable of mobilising adjacent DNA sequences, presenting a very effective way of transferring resistance genes (OCHMAN et al., 2000). As such, these elements mobilise genes that can confer resistance to bacteria, and act as a factor that intensifies the spread of this type of resistance in these organisms (OLIVEIRA; SILVA, 2009). Figure 8 shows a simplified scheme for mobilising a resistance gene into a plasmid via mobile genetic elements.

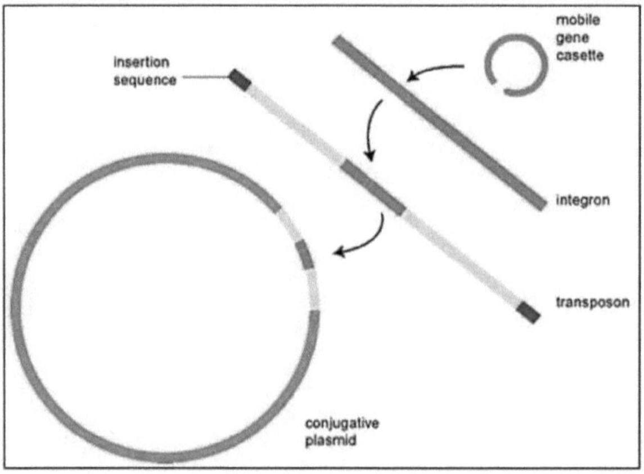

Figura 8- Representation of the plasmid and its mobile genetic elements

Source: Modified from Norman et al, (2009).

Horizontal gene transfer can occur through transformation, transduction and conjugation. Transformation involves the uptake and integration of exogenous DNA into the chromosome, causing recombination by a competent bacterium. This transfer mechanism is more efficient when the bacteria are closely related (OLIVEIRA, SILVA, 2009).

Transduction is a mechanism that occurs through a bacteriophage, which mediates the transfer of DNA from the donor bacterium to the recipient. This transfer takes place when the bacteriophage infects the bacterium and contains the genetic material within its capsid (OLIVEIRA, SILVA, 2009).

Conjugation, the main mechanism for transferring resistance genes, requires direct contact between the donor and recipient cells. This transfer is mediated by a plasmid, which in Gram-negative bacteria expresses sexual pili and allows its transfer to another bacterial cell. In this process, the conjugative plasmid (F factor) of a donor cell (F^+) is replicated and a single strand of this plasmid is transferred via the *pili* to the recipient cell (F^-) where the complementary strand is synthesised (OLIVEIRA, SILVA, 2009).

8.2.9 CTX-M gene

The CTX-M gene codes for a group of enzymes belonging to class "A", according to Ambler's classification, which in the last decade has become the most prevalent group of ESBL enzymes in Enterobacteriaceae. Initially, this gene encoded enzymes capable of preferentially hydrolysing a broad-spectrum cephalosporin (3° generation of cephalosporins), cefotaxime (WANG et al., 2013).

However, the CTX-M gene has undergone several point mutations intensified by the indiscriminate use of antimicrobials that have increased its inhibition potential, leading to the emergence of five genotypes differentiated by their amino acid sequences, these being the CTX-M-1, CTX-M-2, CTX-M-8, CTX-M-9, CTX-M-25 subgroups (FOURNIER et aL, 2013; GINN, PARTRIDGE, 2013).

The name CTX-M came about in reference to the hydrolytic capacity of cefotaxime (acronym: CTX - cefotaxime), and the "M" is a reference to Munich, where it was first described, from an *E. coli* isolated from a patient suffering from otitis media (SILVA, SILVA, OLIVEIRA, 2014).

Phylogenetic analyses suggest that the CTX-M gene arose independently of the other types of ESBLs, from chromosomal genes of *Kluyvera* spp from soil, and was subsequently mobilised into a conjugative plasmid by means of the ISEcpl insertion sequence (CARMO et aL, 2012; GINN, PARTRIDGE, 2013; WELLINGTON et aL, 2013).

Discovered in 1989 almost simultaneously in Europe and South America in an isolate of *E. coli, it rose* to prominence over other enzymes in 2000. There are currently more than 130 enzymes, having been identified mainly in *E. coli* (FOURNIER et al., 2013; GINN, PARTRIDGE, 2013).

These enzymes are distributed throughout the world, with ESBLs being prevalent in Enterobacteriaceae, reported not only in the hospital environment, but also in the community. Their expansion and diversification has proved successful due to their association with transposable elements and integrons, the capture of these elements by conjugative plasmids and their successful transfer between bacteria (RODRÍGUEZ, et aL, 2014).

Although this gene is mainly a plasmid gene, it can also be present in chromosomes. However, in these cases the presence of the gene in chromosomes does not increase its spread, but rather promotes its stabilisation and maintenance in the bacteria (RODRÍGUEZ, et aL, 2014).

8.2.10 Epidemiology of the CTX-M gene

One of the consequences of the generalised increase in the CTX-M gene has been the shift in prescribing from third generation cephalosporins to carbapenems such as imipenem and meropenem as the last treatment option. However, the intensive use of carbapenems has favoured the emergence of carbapenemase enzymes (WELLINGTON et aL, 2013).Wellington et aL (2013) describes CTX-M-15 (a group 1 genotype) as the gene that is widely distributed and commonly recorded throughout the world, having reached prevalence on the Asian continent, in Southern Europe and South America. In Brazil, Enterobacteriaceae are considered to be the main public health problem, since they have a high incidence in clinical isolates in the hospital environment, related to infectious conditions and because they are the main group related to the production of different types of ESBLs (SILVA, 2012).

In the hospital environment, the pathogenic enterobacteria *Klebsiella pneumoniae* and *E. coli* are commonly related to the production of ESBLs and among these, the CTX-M enzymes are the most prevalent, showing wide dissemination. However, several states do not report their presence, since the country does not have a surveillance system for bacterial resistance (SILVA, 2012). Figure 9 shows the distribution of the CTX-M gene subgroups in Enterobacteriaceae in the Brazilian states where they have been analysed and described up to 2012, but the country still has no monitoring programme for the distribution of ESBLs and for controlling the spread of resistant organisms.

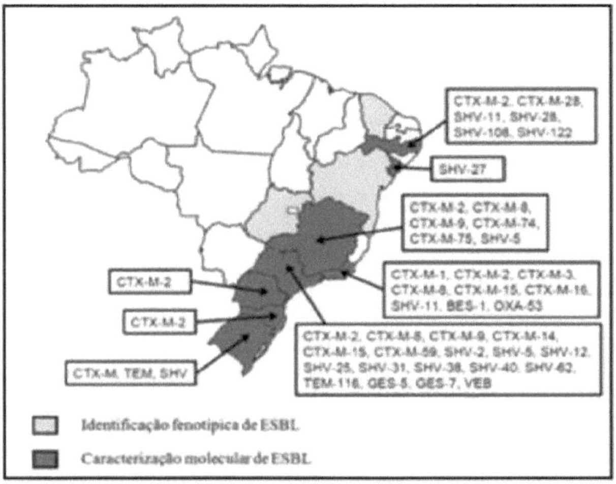

Figura 9- Distribution of the CTX-M gene in Brazil

Source: Silva (2012).

1.2.11 Laboratory identification of bacteria

Since the resistance profile varies greatly between different species and even between individuals of the same species, analyses to identify the species and their resistance/susceptibility profile are of great importance for correct clinical treatment (SILVA, SILVA, OLIVEIRA, 2014).

One of the most widely used methods for identifying genera and even species of certain bacteria at low cost is phenotypic identification using biochemical tests. This technique consists of offering bacteria different nutrients for their growth. The identification of genus and species is carried out through the reactions that occur in different media, such as the degradation of sugars, amino acids, motility, among others (SILVA, SILVA, OLIVEIRA, 2014).

One of the most widely used methods for identifying ESBL-producing strains is the disc diffusion test using the criteria recommended by the *Clinical and Laboratory Standards Institute* (CLSI). This technique consists of applying paper discs impregnated with antimicrobials following a standard of 0.5 on the Mc Farland scale in the middle of a Mueller-Hinton agar plate, 4 mm high in a bacterial suspension (CLSI, 2012).

When the discs are applied to a bacterial suspension, a halo of inhibition appears around the discs in the case of susceptible bacteria. Unlike resistant bacteria in which this halo is not present or is diminished, the latter can be classified as intermediate resistance (AHMED et aL, 2013).A major problem related to the identification of ESBL is the presence of the gene that encodes the enzyme in some microorganisms, but its expression is not detected in the phenotypic tests for its identification. This variation in sensitivity is related to the plasmid in which the gene is inserted, or the promoter region that controls ESBL synthesis (CARMO et aL, 2012).

1.2.12 Identification of the resistance gene

In order to identify resistance genes, effective DNA extraction methods must be used. There are currently a variety of strategies used to extract bacterial DNA, such as chemical, enzymatic and heat shock methods (AHMED, ASGHAR, ELHASSAN, 2014).

From the extracted genetic material, the gene can be amplified using the Polymerase Chain Reaction (PCR) technique. This technique consists of amplifying segments of DNA using specific sequences of oligonucleotides (primers) that are attached to the opposite strands in regions that flank the fragment to be amplified, as well as providing a free 3' OH terminal to start the polymerisation of the segment (CALVETE, CASEIRO, SOUZA, 2015).Visualisation of bands amplified by PCR is possible using the electrophoresis technique, which consists of transmitting an electric current in a vat through a buffer solution for a certain time. The formation of a magnetic field causes the DNA fragments to migrate through an agarose or polyacrylamide gel mesh, separating the molecules by size and charge (FRANCO et al, 2012).

1.2.13 Risk factors for infection with ESBL-producing bacteria

Several factors can lead to infection with ESBL-producing bacteria, including prolonged hospital stays, as well as intensive care units (ICUs) and the use of invasive methods such as mechanical ventilation and intubation, the use of central venous, arterial and urinary catheters, haemodialysis and surgery (GOMIDES et aL, 2014).

According to Ferreira et aL (2011), the problem caused by ESBLs is mainly related to the fact that broad-spectrum antimicrobials such as carbapenems are used in infectious treatments, accelerating the selection mechanism of resistant microorganisms, which leads to prolonged hospital stays and increased treatment costs, intensifying the risk of death.

1.2.14 Control measures

Measures to control resistance are extremely important, as they aim to make the pathogen unviable and thus control the spread of resistance to a certain extent. One of these measures is related to the production and implementation of new antimicrobials or mechanisms that make it possible to invalidate structures that act by intensifying the virulence of the microorganism, such as drugs with direct effects on the outer membrane, making it possible to alter the profile of the microorganism by making it sensitive to the drug (OLIVEIRA, SILVA, 2009; TORTORA, FUNKE, CASE, 2010).

The use of higher concentrations of the antimicrobial in alternation and/or in association with other antimicrobials at an appropriate time and interval tends to overcome the bacteria's inactivation mechanism and thus enable its total destruction in the host (OLIVEIRA, SILVA, 2009; TORTORA, FUNKE, CASE, 2010).

Other measures, such as the use of enzyme synthesis inhibitors that inactivate antimicrobials and simple control of antimicrobial prescriptions by health professionals, as well as correct use by the patient, are extremely important for minimising the mechanisms for the emergence and spread of new types of resistance (OLIVEIRA, SILVA, 2009; TORTORA, FUNKE, CASE, 2010).

CHAPTER 2

OBJECTIVES

Based on studies of bacterial resistance caused by the CTX-M gene, the objectives for this research were defined.

2.1 GENERAL OBJECTIVE

To identify the presence of the CTX-M resistance gene in ESBL-positive bacteria isolated from clinical samples of patients in Roraima between July 2015 and July 2016 using molecular tests.

2.2 SPECIFIC OBJECTIVES

- Identify ESBL-producing strains phenotypically.
- Estimate the major genus/species related to the CTX-M gene among the ESBL-positive strains.
- To trace the antimicrobial resistance/sensitivity profile.
- Correlate phenotypic and molecular data.

CHAPTER 3

MATERIALS AND METHODS

The experimental part of the research was carried out at the Molecular Biology Laboratory (LaBMol) of the Centre for Biodiversity Studies (CBio) at the Federal University of Roraima (UFRR) and at the Central Public Health Laboratory of Roraima (LACEN/RR) from July 2015 to July 2016. This project was approved by the Research Ethics Committee (COEP) on 09/12/2014, opinion no. 935.950.

3.1 SAMPLING

The bacterial samples used in this work came from various biological materials, such as catheter tips, wound secretions, sputum, blood and urine, collected by the clinical staff of the Hospital Geral de Roraima (HGR) and Maternidade Nossa Senhora de Nazaré. The biological samples were collected from patients who had a presumptive diagnosis of infection and were then sent to LACEN/RR for use in this project.

The methodology used for the subsequent identification of the CTX-M gene consisted of the stages of isolation, bacterial culture, biochemical identification and the presence of ESBLs, carried out in collaboration with the team from LACEN's bacteriology department. The subsequent stages of DNA extraction, PCR and electrophoresis to identify the CTX-M gene were carried out at LaBMol/UFRR.

3.2 ISOLATION, BACTERIAL CULTURE AND BIOCHEMICAL IDENTIFICATION.

The techniques described for culture and phenotypic identification by biochemical tests were developed in accordance with the protocol of the National Health Surveillance Agency (ANVISA, 2007).

The medium used to grow the strains was MacConkey Agar for the exclusive growth of Gram-negative bacteria. This medium is used to isolate Gram-negative bacteria because its principle is crystal violet, which inhibits the growth of Gram-positive strains. The bacteria were isolated on this medium by repicking from a disposable sterile plastic loop as shown in (figure 10). They were then incubated at 37°C in an oven for 24 hours.

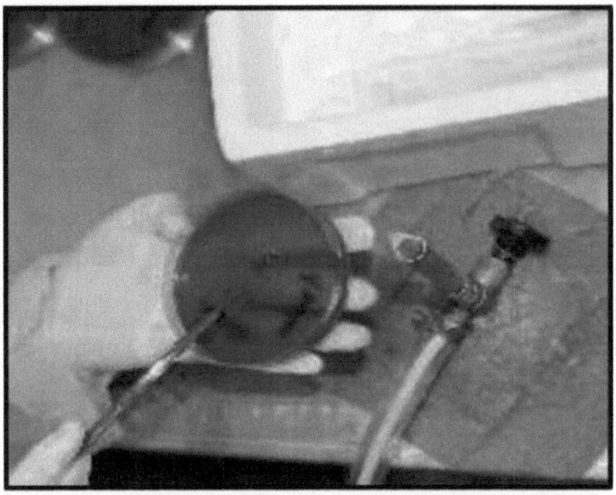

Figura 10- Isolation of Gram-negative bacteria on MacConkey medium

Source: Brandão (2016).

Subsequently, the genus/species of Gram-negative bacteria were identified by means of biochemical tests in different media by analysing the reactions that occurred, or didn't occur, such as fermentation and/or decarboxylation of various biochemical compounds shown in (figure 11) and described below.

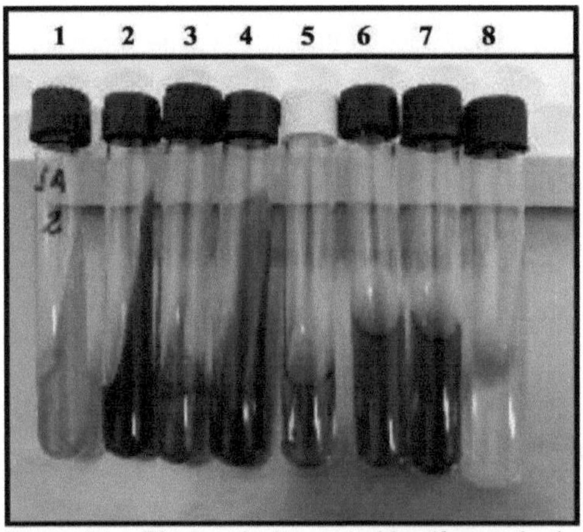

Figura 11- Biochemical tests to identify species/genera

Source: Brandão (2016).

Legend: Order of identification media: 1-Triple Sugar Iron (TSI); 2- Simmons Citrate; 3- Sulphate/Indol/Motility (SIM); 4- Lysine Iron Agar (LIA);
5- Malonate; 6- Ornithine; 7- Arginine and 8- Phenylalanine.

- Triple sugar iron (TSI) - Its main characteristic is the differentiation of Gram negative bacilli based on the fermentation of the sugars present in the medium: glucose, lactose and sucrose. A change in the colour of the medium from red to yellow indicates fermentation.

- Simmons Citrate - Used to differentiate members of the Enterobacteriaceae family and non-fermenters into genus and species, with sodium citrate as the sole carbon source and ammonium dihydrogen phosphate as the sole nitrogen source. The change in colour of the medium from green to blue indicates bacterial growth in response to the use of citrate as a carbon source.
- Sulphate/ Indol/ Motility (SIM) - Allows differentiation of enterobacteria by means of motility, showing diffuse growth outside the puncture point, indicating positive motility. Growth only on the sting line of the medium indicates negative motility.
- Lysine Iron Agar (LIA) - Its purpose is to differentiate micro-organisms based on the decarboxylation and deamination of lysine, the product of which is lysine sulphide.

hydrogen (H2S). A positive result is indicated by a purple colour throughout the tube. A negative result is indicated by a yellowish colour, especially at the bottom of the tube.

-Malonate - Determines the microorganism's ability to use sodium malonate as its sole carbon source, resulting in alkalisation of the medium. A positive result turns the medium blue and a negative result turns it green (unchanged).

-Ornithine Decarboxylase and Arginine Decarboxylase - Decarboxylases are a group of enzymes with a specific substrate, capable of reacting with the carboxyl group of amino acids to form alkaline amines. This reaction, known as decarboxylation, produces carbon dioxide as a secondary product. A positive result is indicated by the colour of the amino acid medium turning purple and cloudy, as amines are formed from the decarboxylation reaction. A negative result remains yellow.

-Phenylalanine - Checks the bacteria's ability to produce phenylpyruvic acid from phenylalanine by enzymatic action. Positive: formation of a greenish colour on the surface of the medium after the addition of ferric chloride. Negative: the medium remains unchanged.

-Urea- Allows the identification of fermenting and non-fermenting Gram negative bacilli through the degradation of urea by the urease enzyme into two ammonia molecules, resulting in an alkalinised medium. The original colour of the medium is straw yellow. A positive interpretation is indicated by the colour of the medium changing to pink.

3.3 PHENOTYPIC IDENTIFICATION OF ESBL

After phenotypic identification, the antibiogram was carried out using the disc diffusion test described by Silva, Silva and Oliveira (2014), following the specifications for interpreting sensitivity results according to the Clinical and Laboratory Standards Institute (CLSI, 2012), in document M100-S22.

In this test, the bacteria previously identified phenotypically were inoculated onto plates containing Mueller-Hinton agar medium. Next, amoxycillin/clavulanic acid was placed in the centre of the plate to inhibit the action of ESBLs. The marker antimicrobials aztreonam, cefotaxime, ceftazidime and cefepime were placed around them at a distance of 20 mm, used for phenotypic identification of the presence/absence of ESBLs by means of the appearance of the synergistic zone (phantom zone) between them and the inhibitor in the centre.

The antimicrobials amikacin, ampicillin/sulbactam, ampicillin, cefazolin, ceftriaxone, cefoxitin, Imipenem, levofloxacin, piperacillin/tazobactam, tetracycline, sulphatrimethoprim, gentamicin, chloramphenicol and cephalexin were distributed on the same plate, totalling 18 antimicrobials tested to check the resistance profile, as shown in the figure below.

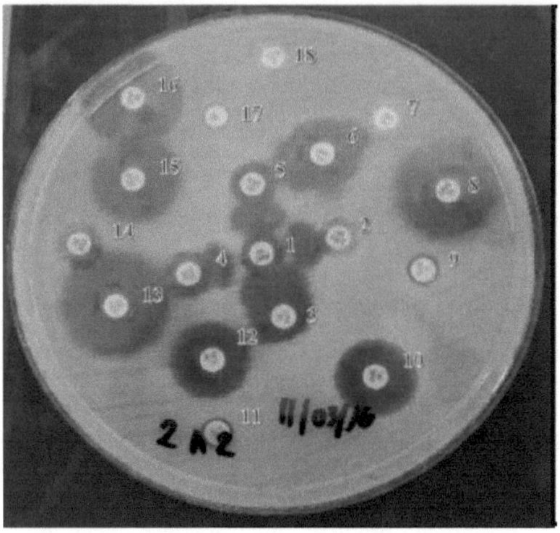

The antimicrobials are shown clockwise from the centre:

1: Amoxicillin + Clavulanate (centre)
2: Aztreonam;
3: Cefotaxime;
4: Ceftazidime;
5: Cefepime;
6: Ceftriaxone;
7: Piperacycline + Tazobactam;
8: Cephalexin;
9: Levofloxacin;
10: Amikacin;
11: Gentamicin;
12: Cefoxitin;
13: Imipenem;
14: Ampicillin/Sulbactam;
15: Chloramphenicol;
16: Ertapenem;
17: Sulfatrimethoprim
18: Tetracycline.

Figura 12- Antibiogram for identification of ESBL enzyme and resistance/susceptibility profile
Source: Brandão (2016).

Afterwards, the plate on which the discs were distributed was incubated for 18 to 20 hours at 35°C. At the end of the process, the presence or absence of a zone with distortion of the halo (ghost zone) was checked. If a ghost zone appeared between the ESBL inhibitor in the centre of the plate and the marker discs positioned around it, the test was interpreted as a positive result for the presence of the ESBL enzyme.

3.4 MOLECULAR IDENTIFICATION OF THE CTX-M GENE

A heat shock DNA extraction technique proposed by Chapman et al. (2001) was used, which consists of lysing the bacterial wall and exposing the genetic material by boiling, followed by a PCR reaction of the samples to analyse the CTX-M gene.

3.5 DNA EXTRACTION

In the heat shock DNA extraction proposed by Chapman et al. (2001). The bacteria identified as ESBL positive were seeded for 24 hours at 37°C in BHI *(Brain Heart Infusion)* liquid medium. The culture broth containing the bacterial growth was then transferred to a 1.5 ml microtube. The samples were centrifuged at 6,500 rpm for 10 minutes. The supernatant was discarded and the precipitate resuspended in 400 ul of ultrapure water. The sample was then subjected to the boiling extraction method, as shown in the figure, by submitting it to a water bath at 100°C for 10 minutes. It was then immediately cooled in a freezer to -80°C and kept at 20°C until processing.

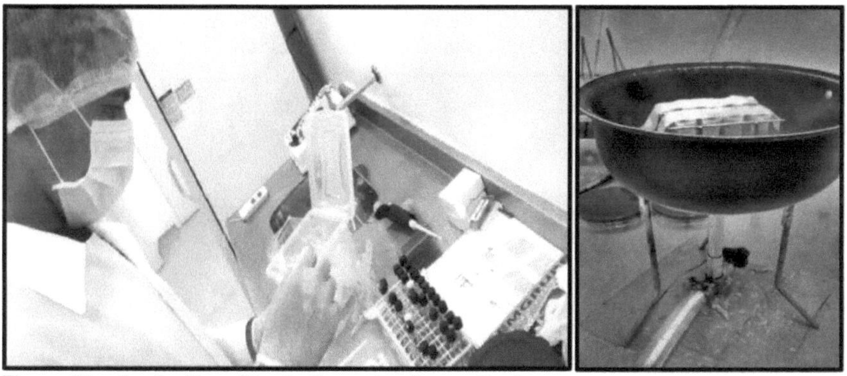

Figura 13- Application of the boiling extraction method

3.6 POLYMERASE CHAIN REACTION (PCR)

In the Polymerase Chain Reaction (PCR), specific primers were used, as shown in Table 1, to delimit the plasmid DNA region of approximately 593 base pairs (bp) that encodes the CTX-M enzyme, following the methods of Ferreira et al. (2011).

Table 1 - PCR primer sequence

Genes	Oligonucleotide sequence (5'- 3')	Product (pb)
CTX-M-F	ATGTGCAGYACCAGTAARGT	Approximately 593 bp.
CTX-M-R	TGGGTRAARTARGTSACCAGA	

Source: Adapted from Ferreira et al. (2011).

The following volumes of reagents were used for the PCR reaction: 0.5 µl/microtube dNTPs, 0.5 µl/microtube CTX-M-F primer at 25 µM and 0.5 µl/microtube CTX-M-R primer at 25 µM, 2.5 µl/microtube 10x buffer, 0.75 µl/microtube magnesium chloride (MgCh), 0.5 µl/microtube *Taq* DNA polymerase (Ludwig), for 1 µl/microtube of sample and enough nuclease-free ultrapure water for a total reaction of 25ul/microtube.In a thermal cycler (PXE 0.2 Thermal Cycler), the samples were subjected to 1 cycle for 7 minutes at 94°C, 35 cycles for 50 seconds at 94°C, 40 seconds at 50°C, 1 minute at 72°C; 1 cycle for 5 minutes at 72°C, resulting in *amplicons* with sequences homologous to the DNA of origin.

3. 7ELECTROPHORESIS

Amplification of the CTX-M gene was carried out by electrophoresis prepared on agarose gel diluted in a concentration of 1.5% of the volume of TBE (Tris Borate EDTA) buffer at 0.5. After the gel was completely polymerised, a *mix* consisting of 5 µl of PCR, 2 µl of 10% bromophenol blue and 0.7 µl of *Blue Green* dye (LGC) was introduced. They were then subjected to an electric field for 30 minutes at 100 V (FRANCO et al., 2012). As shown in (figure 14) below.

Figura 14- Application of the *mix* for the 1.5% agarose gel electrophoresis reaction.

The molecular weight marker (Ludwig Biotec) used comprises DNA fragments simulating a chain of 100 bp to 2,000 bp, allowing comparison of the size of the amplified fragments of approximately 593 bp. (FRANCO et al, 2012). The fragments were then visualised on the MiniBis photodocumentator, with the *Klebsiella pneumoniae* sample (17CTXLaBMol/UFRR) sequenced by the UFRR Molecular Biology Laboratory as a positive control.

3.8 STATISTICAL ANALYSIS

For the statistical interpretation of the data, the Chi-squared test was used for the relationship between the bacteria involved in the study and the frequency of the CTX-M gene.

CHAPTER 4

RESULTS AND DISCUSSION

Between July 2015 and July 2016, 72 strains (Table 1) were positive for ESBLs using the disc diffusion test. The biochemical series identification method showed that *Klebsiella sp.* with 37.5% of the representatives, was prevalent for the production of extended-spectrum beta-lactamase enzymes (ESBLs). This was followed by *Pseudomonas sp.* 16.6%, BGN 15.3%, *Enterobacter sp.* 13.9%, *E. coli* with 12.5%, *Serratia sp.* 2.8% and *Proteus Vulgaris* with 1.4% of the representatives.

Table 1- Identification of the strains phenotypically confirmed as ESBL producers

Micro-organisms	Phenotypically identified strains	%
***Klebsiella* sp.**	27	37,5
***Pseudomonas* sp.**	12	16,6
BNG	11	15,3
***Enterobacter* sp.**	10	13,9
Escherichia coli	9	12,5
***Serratia* sp.**	2	2,8
Proteus Vulgaris	1	1,4
Total	**72**	**100**

(n = 72, $x^2 = 42.722$, $p < 0.0001$)

According to Carmo et al. (2012), strains containing genes encoding resistance enzymes may or may not express them in phenotypic tests, depending on the plasmid in which the gene is inserted or the presence of a promoter region necessary for synthesising the enzyme. Since degradation of the beta-lactam ring occurs slowly, it may not be visualised *in in vitro* tests.

Another factor that may make it impossible to identify this phenotypic profile is related to the quantity of the inoculum, which may be in low concentrations, as well as the enzyme, making it difficult to visualise the formation of the growth inhibition halo.

These statements may overestimate the sampling of ESBL-carrying strains during the period of this research and justify the sensitivity of strain no. 5 - *Klebsiella ozaenae* (shown in table 4 and figure 17) to cefotaxime, since there was molecular confirmation of the presence of the CTX-M gene in this strain and the main phenotypic characteristic of the presence of this gene is the hydrolysis of this antimicrobial, leading us to infer that in this strain the expression of the gene was silenced.

Farinas and Martínez-Martínez (2013) describe *K. pneumoniae* and *E. coli* as the most important *Enterobacteraceae* species in terms of ESBL enzyme production, especially in intensive care units, surgical and neonatal areas. This is also shown by Mshana et al. (2009), in a study carried out in Tanzania using clinical strains, which disagrees with what was observed in this study, where the prevalence was of *Klebsiella sp.* with 37.5 per cent of the representatives and *Pseudomonas sp.* 16.6 per cent for the presence of ESBLs.

This research aimed to identify resistance in clinical isolates, however, data presented by Silva (2012) showed that this dissemination is not restricted to the hospital environment and can be found in isolates of environmental and community origin and in food of animal origin.

Among the 72 ESBL-positive strains, the presence of the CTX-M resistance gene was observed by PCR in 29% of the strains. Figure 15 shows the 593 bp fragments amplified in the PCR for the CTX-M gene, visualised on a 1.5% agarose gel.

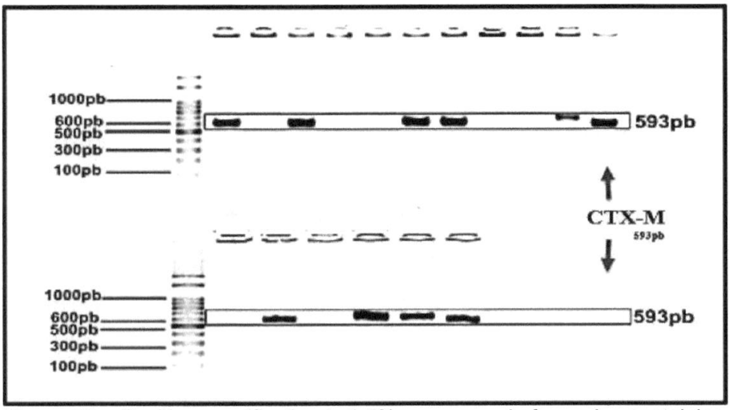

Figura 15- Positive amplification in 1.5% agarose gel of samples containing CTX genes by PCR

The data in Table 2 shows that the genus *Klebsiella sp.* with 13 representatives (61.9%) is the predominant carrier of this gene, followed by *Enterobacter sp.* with 3 (14.3%), Gram-negative non-glucose fermenting bacillus with 2 (9.5%), *Escherichia coli with* 2 (9.5%) and *Pseudomonas sp.* with 1, (4.8%) representatives.

Table 2- Distribution of the CTX-M gene in clinical strains

Bacteria	N°	By species (%)	By gender (%)
Klebsiella ozaenae	**6**	28,58	61,9
Klebsiella sp.	5	23,8	
Klebsiella rhynoscleromatis	1	4,8	
Klebsiella pneumoniae	1	4,8	
Enterobacter sp.	2	9,5	14,3
Enterobacter cloacae	1	4,8	
BGN	2	9,5	9,5
Escherichia coli	2	9,5	9,5
Pseudomonas sp.	1	4,8	4,8

Total	21	100%	100%

(X^2 = 29,851, $p < 0.0005$).

The *Klebsiella* genus was prevalent among the strains identified phenotypically in this study. These results agree with those described by Lima, et al. (2014); Moreira (2014); Silva (2012), pointing to them as the main group related to hospital outbreaks.

According to Moreira (2014); Farinas, Martínez-Martínez (2013), the microbiological diagnosis of multidrug-resistant enterobacteria infections presents particular difficulties, since these microorganisms grow well on conventional media. In addition, reading the metabolic profile of different species of the same genus can often be very similar and make identification at the species level difficult, since the determination of microbial susceptibility follows a pre-established standard.

These statements may justify the identification down to genus of some strains in this research, leading us to suggest alternative identification methods to confirm the species.

Since beta-lactamases have been recognised as the main cause of antimicrobial resistance in members of the *Enterobacteraceae* family, Castanheira (2013), in a study carried out in 26 hospitals in the United States on the presence of different resistance genes from 157 ESBL-positive isolates, observed the prevalence of the CTX-M gene in 46.6% of the sample, being mainly related to *E. coli* strains and not to the Klebsiella genus, as demonstrated in this study. According to the Antimicrobial Surveillance Programme (SENTRY), the rates of CTX-M producing strains seem to be increasing in US hospitals (26.6% in 2007 to 43.8% in 2010).

Stefani et aL (2014), searching for EBLSs in pigs and cattle in northern Italy, showed that of the 56 samples analysed, 20 (36%) were positive for ESBL production using the double-disc synergy test. Of these, 15 (75 per cent) were *E. coli* and 5 (25 per cent) of the species were *K. ozaenae,* showing prevalence of the CTX-M gene as the most prevalent ESBL type. Highlighting the role of consumer animals in the spread of ESBLs in Enterobacteriaceae, and the risk of these micro-organisms reaching humans through the food chain.

Dropa (2016), analysing wastewater from treatment plants in the city of São Paulo, Brazil, observed the prevalence of the CTX-M gene belonging to the CTX-8 and CTX-M 15 variants in *Klebsiella sp. and E. coli* species.

This study presented a single strain of *Pseudomonas sp.* This genus includes species that are widely distributed in different environments, especially in humans. The *P. aeruginosa species* is more important because it is the main one related to serious infections with high morbidity and mortality in immunosuppressed patients (FARINAS, MARTÍNEZ-MARTÍNEZ, 2013).

The *Pseudomonas sp.* strain observed in this study was sensitive only to imipenem, levofloxacin, piperacillin/tazobactam and gentamicin, representing 22.2 per cent of the antimicrobials tested. According to Farinas and Martínez-Martínez (2013), the entire genus, but especially the *P. aeruginosa* species, is naturally resistant to most second and third generation cephalosporins.

The most prevalent clinical material analysed for the detection of CTX-M came from samples of urine, blood, catheter tips, wound secretions and tracheal secretions. These data are shown in Table 3. We observed similar results regarding the presence of strains carrying the CTX-M resistance gene among the different biological materials presented, demonstrating that the strains carrying the gene that codes for the CTX-M enzyme in this study are involved in the different types of infection.

Table 3- Prevalence of clinical materials collected at the Roraima General Hospital

Clinical Material	N°	%
Urine	5	23,8
Blood	4	19,05
Catheter tip	4	19,05
Wound secretion	4	19,05
Tracheal secretion	4	19,05
Total	**21**	**100**

In the hospital environment, urinary tract infections are among the most common. The incidence of ESBL in this type of infection has increased in recent years, with the main risk factor being the use of a vesical catheter. Patients with a history of previous hospitalisation, previous urogenital surgery and advanced age also act as risk factors for infection by ESBL bacteria in the urinary tract (AZAP, et aL 2010; LEE, 2010). According to Sfair et aL (2014), the risk of developing a urinary tract infection related to catheter use increases by 5% for every day of use.

Azap et al (2010), in their research carried out in Turkey with patients diagnosed with urinary tract infection, observed that *E. coli* was mainly related to this type of infection and in analysing the resistance gene of these strains, the prevalence of CTX-M was observed.

In order to describe the resistance/susceptibility profile of the strains involved in this research, 18 antimicrobials were placed in Petri dishes in a bacterial suspension: amikacin, ampicillin/sulbactam, ampicillin, aztreonam, cefazolin, ceftadizime, ceftriaxone, cefotaxime, cefoxitin, cefepime, imipenem, levofloxacin, piperacillin/tazobactam, tetracycline, sulphatrimethoprim, gentamicin, chloramphenicol, cephalexin, around the ESBL inhibitor amoxillin/clavulanic acid.

Table 4 shows the antimicrobials that were susceptible to inhibiting the growth of the 21 species carrying the CTX-M gene. There was greater susceptibility, i.e. a positive response to treatment with piperacillin + tazobactam with 90.5% of the strains showing sensitivity to this antimicrobial, followed by imipenem 85.7%, levofloxacin 47.6%, amikacin and chloramphenicol with 42.8% of the sensitive strains.

Tamma et al. (2015), reported the efficacy of the betalactamase inhibitor tazobactam associated with piperacillin and imipenem, as a therapy alone or in combination with other antimicrobials for the treatment of ESBL infections. This statement agrees with what was observed in this study, in which the combination of piperacillin + tazobactam was 90.5% effective, followed by imipenem with 85.7% effectiveness in inhibiting microbial growth.

According to Perez and Bonomo (2012), the treatment of ESBL-producing organisms with piperacillin associated with tazobactam exerts great selective pressure on strains of *Pseudomonas aeruginosa* and that although the addition of tazobactam appears to reduce the hydrolysis effect of the enzymes, the activity of tazobactam is diminished when a high concentration of bacteria is present.

Gudiol et al. (2010), when analysing sensitivity rates to betalactam antimicrobials, observed that all the strains remained completely sensitive to imipenem and amikacin. These results disagree with those found in this study, in which imipenem and amikacin showed 85.7 and 42.8 per cent susceptibility, respectively.

Table 4- Sensitivity profile by species

N°	SPECIES	Susceptible antimicrobials
1	***Klebsiella ozaenae***	**AMI, IMP, PPT, TETRA, GEN**
2	***Klebsiella ozaenae***	**AMI, CEFAZO, IMP, PPT, TETRA, CLO**
3	***Klebsiella ozaenae***	**AMI, IMP, LEVO, PPT**
4	***Klebsiella ozaenae***	**AMI, IMP, LEVO, PPT**
5	***Klebsiella ozaenae***	**AMI, ATM, CEFAZO, CRO, CEFOT, IMP, LEVO, PPT, CLO**
6	***Klebsiella ozaenae***	**IMP, LEVO, PPT**
7	***Klebsiella sp.***	**PPT**
8	***Klebsiella sp.***	**AMI, IMP, PPT, TETRA**
9	***Klebsiella sp.***	**AMI, IMP, PPT, CLO**
10	***Klebsiella sp.***	**IMP, PPT, CLO**
11	***Klebsiella sp.***	**IMP, PPT**
12	***Klebsiella rhynoscleromatis***	**AMI, IMP, PPT, CLO**
13	***Klebsiella pneumoniae***	**AMI, IMP, LEVO, PPT**
14	***Enterobacter sp.***	**IMP, LEVO, PPT, CLO**
15	***Enterobacter sp.***	**IMP, LEVO, PPT**
16	***Enterobacter cloacae***	**IMP, LEVO, PPT, GEN, CLO**
17	**BGN**	**ABSENCE**
18	**BGN**	**ABSENCE**
19	***Escherichia coli***	**IMP, LEVO, PPT, TETRA, CLO**
20	***Escherichla coli***	**IMP, PPT, TETRA, CLO**
21	***Pseudomonas sp.***	**IMP, LEVO, PPT, GEN**

Key: AMI: Amikacin; ATM: Aztreonam; CEFAZO: Cefazolin; CRO: Ceftriaxone; CEFOT: Cefotaxime; CEFEP: Cefepime; IMP: Imipenem; LEVO: Levofloxacin; PPT: Piperacillin/Tazobactam; TETRA: Tetracycline; GEN: Gentamicin; CLO: Chloramphenicol; CFL: Cephalexin; BGN: Gram-negative non-glucose fermenting bacillus.

When analysing the resistance/susceptibility profile of the strains carrying the CTX-M gene in this study, high levels of resistance were observed for most of the antimicrobials tested. 100% of the strains showed resistance to ampicillin/sulbactam, ampicillin, ceftadizime, cefoxitin, cefepime, sulphatrimethropime and cephalexin, followed by aztreonam, ceftriaxone and cefotaxime with

95.2%, cefazolin 90.5%, gentamicin 85.7% and tetracycline 76.2%. These results, presented in raw values in figure 12, demonstrate the limited treatment options for infections and the high resistance profile of the ESBL CTX-M strains.

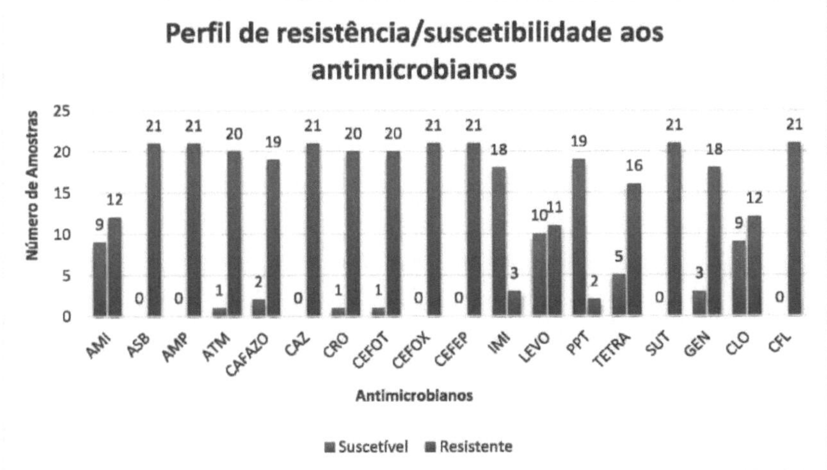

Figura 16- Antimicrobial resistance/susceptibility profile of strains carrying the CTX-M gene

Legend: AMI: Amikacin; ASB: Ampicillin/Sulbactam; AMP: Ampicillin; ATM: Aztreonam; CEFAZO: Cefazolin; CAZ: Ceftadizime; CRO: Ceftriaxone; CEFOT: Cefotaxime; CEFOX: Cefoxitin; CEFEP: Cefepime; IMP: Imipenem; LEVO: Levofloxacin; PPT: Piperacillin/Tazobactam; TETRA: Tetracycline; SUT: Sulfatrimethropime; GEN: Gentamicin; CLO: Chloramphenicol; CFL: Cephalexin.

According to Shaikh (2015), the detection of ESBLs and their resistance/susceptibility profile is of great importance, as it makes it possible to minimise the spread of these bacteria and help select the most appropriate antimicrobial.

The susceptibility of amikacin corresponded to 42.8 per cent efficacy in this study. Salvatore and Resman-Targoff (2015) demonstrated the efficacy of monotherapy with this antimicrobial in the treatment of cystitis and bacterial pyelonephritis in infections with ESBL-producing bacteria. According to the authors, a benefit of amikacin is that high urinary concentrations are achieved, providing between 94-98% of the drug unchanged, eliminating it through the urine within 24 hours.

The results of this research indicated two representatives of Gram-negative non-glucose fermenting bacilli (GNB), and one of *Pseudomona sp.,* both not belonging to the *Enterobactereaceae* family. High resistance to the antimicrobials tested was observed among these strains. The BGN showed 100 per cent resistance to the antimicrobials tested and *Pseudomona sp. to* 77.8 per cent. In a study of ESBLs carried out by Lago et al. (2010) in a hospital in Passo Fundo-RS, it was observed that in addition to strains of *E. coli and Klebsiella sp.* being mainly related to this resistance mechanism, *non-Enterobacteraceae* species are widely disseminated with this resistance mechanism.

Shaikh (2015) describes in his research the significant increase in resistance worldwide due to the production of ESBL enzymes and points to the CTX-M variant as its main representative. The results obtained in his research show high resistance to the antimicrobials Gentamicin, tetracycline and ceftriaxone in clinical strains from Nigeria, which agrees with the results of this research.

Azap et al. (2010), in a study carried out in Turkey on clinical samples isolated from the urinary tract, found that 57% of their samples were resistant to gentamicin, a result different from that observed in this study, in which 85.7% of the strains were resistant to gentamicin.

The results presented demonstrate the high resistance profile of ESBL-positive strains that possess the CTX-M gene, which leads to restrictions in treatment options for individuals with infections. Measures to control the spread of resistance are proposed, such as the production and implementation of new antimicrobials, the use of higher concentrations of these drugs in alternation and/or in association with others at an appropriate time and interval, the use of enzyme synthesis inhibitors and simple control in the prescription and appropriate use of drugs.

CHAPTER 5

CONCLUSION

- The research carried out showed a high rate of positive ESBL strains for the period in which this analysis was carried out, with a predominance of *Klebsiella* sp.
- The CTX-M resistance gene was present in 29% of the strains identified as ESBL-positive. Of this total, *Klebsiella sp.* was predominant in 61.9% of the samples, followed by *Enterobacter sp.* with 14.3%, Gram Negative Non-Glucose Fermenting Bacillus with 9.5%, *Escherichia coli* with 9.5% and *Pseudomonas sp. with* 4.8%.
- The strains carrying the CTX-M resistance gene showed high levels of resistance to most of the antimicrobials tested. They were most sensitive to treatment with piperacillin + tazobactam (90.5%), followed by imipenem (85.7%), levofloxacin (47.6%), amikacin and chloramphenicol (42.8%). These results suggest that these antimicrobials should be used as the therapy of choice for patients with suspected infections caused by ESBL strains. However, the liberal use of these antimicrobials may result in the emergence of resistance to these drugs, severely limiting future treatment options.
- Using the polymerase chain reaction, it was possible to confirm that bacteria with ESBLs had the CTX-M resistance gene, as well as demonstrating that their spread is intensified in the hospital environment, associated with different types of infection based on their presence in different clinical materials. This leads us to suggest that this spread may be directly related to the widespread use of antimicrobials and their inappropriate use.
- The high number of ESBL-positive strains and the resistance/susceptibility profile of these strains to the antimicrobials tested leads us to suggest the presence of other resistance genes in the hospital environment in the state of Roraima, whether or not they are associated with the CTX-M gene.

CHAPTER 6

REFERENCES

AHMED, O. I. et al. Detection of bla SHV and bla CTX-M genes in ESBL producing *Klebsiella pneumoniae* isolated from Egyptian patients with suspected nosocomial infections. **Egyptian Journal of Medical Human Genetics,** v. 14, n. 3, p. 277-283, 2013.

AHMED, O. B.; ASGHAR, A. H.; ELHASSAN, M. Comparison of three DNA extraction methods for polymerase chain reaction (PCR) analysis of bacterial genomic DNA. **African Journal of Microbiology Research,** v. 8, n. 6, p. 598- 602, feb. 2014.

ALMEIDA, M. G. C. et al. Sensitivity profile to antimicrobials in *Pseudomonas aeruginosa of* hospital and outpatient origin from public and private laboratories in Belém, state of Pará. **Belém, state of Pará. RBAC,** v. 44, n.1, p. 44-49, 2012.

AMBLER, R. P. The structure of P-lactamases. **Philosophical Transactions of the Royal Society B: Biological Sciences,** v. 289, n. 1036, p. 321-331, May 1980.

ANTONIO, N. S. et al. Mechanisms of bacterial resistance. **Rev. Cient. Elet. Med.** Vet, v. 12, n. 7, p. 1-4, 2009.

ARAÚJO, E. A. et al. Colloidal aspects of microorganism adhesion. **Chem. Nova,** v. 33, n. 9, p. 1940-1948, 2010.

AZAP, Õ. K. et al. Risk factors for extended-spectrum p-lactamase positivity in uropathogenic Escherichia coli isolated from community-acquired urinary tract infections. **Clinicai microbiology and infection,** v. 16, n. 2, p. 147-151,2010.

BEVERIDGE, T. J. Gram-negative cell wall structures and their derived membrane vesicles. **Journal of bacteriology,** v. 181, n. 16, p. 4725-4733, 1999

BLAIR, J. M. A et al. Molecular mechanisms of antibiotic resistance. **Nature Reviews Microbiology,** v. 13, n. 1, p. 42-51, 2015.

BROLUND, A. et al. Plasmidome-analysis of ESBL-producing *Escherichia coli* using conventional typing and high-throughput sequencing. **Public Library Of Science,** v. 8, n. 6, p. 657-693, May/June 2013.

BRAZIL. National Health Surveillance Agency (ANVISA). Clinical Microbiology for Healthcare-Related Infection Control. **Module 6: Detection and identification of bacteria of medical importance.** Brasília: ANVISA, 2013. 150p . Available at : <https://spdbcfmusp.files.wordpress.com/2014/09/iras_modulodeteccaobacterias.pdf>. Accessed on: 15 February 2017.

BRAZIL. National Health Surveillance Agency (ANVISA). Management of Investigation and Prevention of Infections and Adverse Events (Gipea) and General Management of Technology in Health Services (GGTES). **Investigation and Control of Multidrug-Resistant Bacteria.** Brasília: ANVISA, 2007. 21 p. Available at: <http://www.anvisa.gov.br/servicosaude/controle/reniss/manual%20 controle bacteri as.pdf>.

Accessed on: 23 May 2017.

BRANDÃO, M. B. S. **Biodiversity of Extended Spectrum Betalactamase-producing strains in effluents from a public hospital in Boa Vista-RR.**
2016. 80p. Thesis Qualification - Postgraduate Programme in Biodiversity, Federal University of Roraima, Boa Vista, 2016.

BUSH, K. Bench-to-bedside review: The role of P-lactamases in antibiotic-resistant Gram-negative infections. **Critical Care,** v. 14, n. 3, p. 224, 2010.

CALVETE, C. L.; CASEIRO, M. M.; SOUZA, C. B. Biotechnology: bacterial transformation by heat shock method. **UNILUS Ensino e Pesquisa,** v. 12, n. 26, p. 41-53, jan./mar. 2015.

CARTAXO R. S. et al. Mechanisms of resistance and detection of betalactamases. **Journal of Health Sciences,** v. 7, n. 1, p. 59-63, 2015.

CARMO S. D. et al. Detection of extended-spectrum betalactamases (ESBL) in uropathogenic *Escherichia coli* isolates (UPECS) from community patients. **Revista de Patologia Tropical,** v.41, n.4, p.419-426, Oct./Dec. 2012.

CASTANHEIRA, M. et al. Prevalence of P-lactamase encoding genes among Enterobacteriaceae bacteremia isolates collected in 26 USA hospitals: Report from the SENTRY Antimicrobial Surveillance Program (2010). **Antimicrobial agents and chemotherapy,** p. AAC. 02252-12, 2013.

CHAPMAN, P. A. et al. Comparison of culture, PCR and immunoassays for detecting *Escherichia coli* 0157 following enrichment culture and immunomagnetic separation performed on naturally contaminated raw meat products. **International Journal of Food Microbiology,** v. 68, n. 1, p. 11-20, Aug. 2001.

CLSL Performance Standards for Antimicrobial Susceptibility Testing- Twenty Second Informational Supplement. M100-S22. **Clinical Laboratory and Standards Institute,** v.32, n.3, Jan. 2012.

CRUZ-MACHADO, S. S. Lipopolysaccharide (LPS): activator and regulator of gene transcription via transcription factor NFKB. **Revista da Biologia,** v. 4, p. 40-43, Oct./Jan. 2010.

DALMARCO, E. M.; BLATT, S. L.; CÓRDOVA, C. M. M. Laboratory Identification of extended spectrum P-lactamases - A review. **Brazilian Journal of Clinical Analyses,** v. 38, n. 3, p. 171-177, 2006.

DANTAS, G; SOMMER, M. O. A. Context matters - the complex interplay between resistome genotypes and resistance phenotypes. **Current opinion in microbiology,** v. 15, n. 5, p. 577-582, 2012.

DROPA, M. et al. Genetic background of novel sequence types of CTX-M-8-and CTX-M-15-producing Escherichia coli and Klebsiella pneumoniae from public wastewater treatment plants in São Paulo, Brazil. **Environmental Science and pollution research international,** v. 23, n. 5, p. 4953, 2016.

FARINAS, M. C.; MARTÍNEZ-MARTÍNEZ, L. Infections caused by multidrug-resistant gram-negative bacteria: enterobacteria, *Pseudomonas aeruginosa, Acinetobacter baumannii* and other non-fermenting gram-negative bacilli. **Enfermedades Infecciosas y Microbiología Clínica,** v. 31, n. 6, p. 402-409, 2013.

FERREIRA, C. M. et al. Extended-spectrum beta-lactamase-producing bacteria isolated from haematologic patients in Manaus, State of Amazonas, Brazil. **Brazilian Journal of Microbiology,** v. 42, n. 3, p. 1076-1084, 2011.

FOURNIER, D. et al. Alternatives to carbapenems in ESBL-producing Escherichia coli infections. **Médecine et maladies infectieuses,** v. 43, n. 2, p. 62-66, 2013.

FRANCO, A. T. M. et al. Comparison between classical biochemical test and Polymerase Chain Reaction (PCR) method for identification of *Enterococcus faecalis* strains isolated from the oral cavity. ***SALUSVITA,*** Bauru, v. 31, n. 3, p. 191-202, Oct./Dec. 2012

GINN, A. N.; PARTRIDGE, S. R. Generation of the b/aCTX-M-116 Hybrid Gene. **Antimicrobial Agents and Chemotherapy,** v. 57, n. 8, p. 4091-4091, 2013.

GOMIDES, M. D. A. et al. Gram-negative bacteria resistant to carbapenems: analysis of 79 cases from an adult intensive care unit. **Enciclopédia Biosfera,** v. 10, p. 149-57, 2014.

GUDIOL, C. et al. Bacteraemia due to extended-spectrum b-lactamase-producing *Escherichia coli* (ESBL-EC) in cancer patients: clinical features, risk factors, molecular epidemiology and outcome. **Journal of Antimicrob Chemother,** v.65, p.333-341, 2010.

KORZENIEWSKA, E.; HARNISZ, M. Extended-spectrum beta-lactamase (ESBL)-positive Enterobacteriaceae in municipal sewage and their emission to the environment. **Journal of Environmental Management,** v. 128, n. 1, p. 904-911, 2013.

LAGO, A.; et al. ESBL-producing Enterobacteriaceae in Passo Fundo, State of Rio Grande do Sul, Brazil. **Rev. Soc. Bras. Med. Trop,** v. 43, n. 4, p. 430-434, 2010.

LEE, D. S.; LEE, C. B.; LEE, S. J. Prevalence and riskfactors for extended spectrum beta-lactamase-producing uropathogens in patients with urinary tract infection. **Korean Journal of urology,** v. 51, n. 7, p. 492-497, 2010.

LIMA, M. R. S. et al. Intervention in an outbreak of expanded-spectrum betalactamase (ESBL)-producing Klebsiella pneumoniae in a neonatal intensive care unit in Teresina, Piauí, 2010-2011.**Epidemiologia e Serviços de Saúde,** v. 23, n. 1, p. 177-182, 2014.

MACHADO, P. RL et al. Immune response mechanisms to infections. **An Bras Dermatol,** v. 79, n. 6, p. 647-664, 2004.

MARTINEZ, J. L. Environmental pollution by antibiotics and by antibiotic resistance determinants. **Environmental pollution,** v. 157, n. 11, p. 2893-2902, 2009.

MEYER, G.; PICOLI, S. U. Betalactamase phenotypes in *Klebsiella pneumoniae* from an emergency hospital in Porto Alegre. **Brazilian Journal of Pathology and Medicine,** v. 47, n. 1, p. 25-31, feb. 2011.

MOREIRA, V. C.; FREIRE, D. Klebsiella pneumoniae and its resistance to antibiotics, **accessed on: March** 2014.

MSHANA, Stephen E. et al. Prevalence of multiresistant gram-negative organisms in a tertiary hospital in Mwanza, Tanzania. **BMC research notes,** v. 2, n. 1, p. 49, 2009.

NEVES, P. R. et al. Multidrug-resistant Pseudomonas aeruginosa: an endemic problem in Brazil. **Brazilian Journal of Pathology and Laboratory Medicine,** v. 47, n. 4, p. 409-420, 2011.

NORMAN, A.; HANSEN, L. H.; S0RENSEN, S. J. Conjugative plasmids: vessels of the communal gene pool. **Philosophical Transactions of the Royal Society B: Biological Sciences,** v. 364, n. 1527, p. 2275-2289, 2009.

OCHMAN, H.; LAWRENCE, J. G.; GROISMAN, E. A. Lateral gene transfer and the nature of bacterial innovation. **Nature,** v. 405, n. 6784, p. 299-304, 2000.

OLIVEIRA, A. C.; SILVA, R. S. Challenges of health care in the face of bacterial resistance: a review. **Revista eletrónica de enfermagem,** v. 10, n. 1, p. 189-197, 2009.

ÕZTÚRK, H.; OZKIRIMLI, E.; ÕZGUR, A. Classification of Beta-Lactamases and Penicillin Binding Proteins Using Ligand-Centric Network Models. **Public Library of Science,** v. 10, n. 2, p. 1-23, 2015.

PEREZ, F.; BONOMO, R. A. Can We Really Use B-Lactam/B-Lactam Inhibitor Combinations for the Treatment of Infections Caused by Extended-Spectrum B- Lactamase-Producing Bacteria? **Clinicai Infectious Diseases,** v. 54, n. 2, p. 175- 177, 2012.

RODRÍGUEZ, I. et al. Chromosomal localisation of bla CTX-M genes in clinical isolates of Escherichia coli from Germany, The Netherlands and the UK. **International Journal of antimicrobial agents,** v. 43, n. 6, p. 553-557, 2014.

SALVATORE, D. J.; RESMAN-TARGOFF, B. H. Treatment Options for Urinary Tract Infections Caused by Extended-Spectrum B-Lactamase-Producing *Escherichia coli* and *Klebsiella pneumoniae.* **J. Academic Hospital Medicine,** v. 7, n. 1,2015.

SFAIR, S. et al. Risk factors associated with nosocomial urinary tract infection by extended-spectrum betalactamases. **Journal of Infection Control,** v. 3, n. 2, 2014.

SHAIKH, S. et al. Antibiotic resistance and extended spectrum beta-lactamases: Types, epidemiology and treatment. **Saudi Journal of biological Sciences,** v. 22, n. 1, p. 90-101,2015.

SILHAVY, T. J.; KAHNE, D.; WALKER, S. The bacterial cell envelope. **Cold Spring Harbor perspectives in biology,** v. 2, n. 5, p. a000414, 2010.

SILVA, K. C.; LINCOPAN, N. Epidemiology of extended-spectrum betalactamases in Brazil: clinical impact and implications for agribusiness. **J. bras. patol. med. lab,** v. 48, p. 91-99, 2012.

SILVA, R. C. G.; SILVA, A. C. O.; OLIVEIRA, S. R. Microbial resistance and frequency of extended-spectrum beta-lactamase (ESBL) in isolates from blood cultures. **Jornal Brasileiro de Patologia e Medicina Laboratorial,** v. 50, n. 6, p. 421-427, Rio de Janeiro, Dec. 2014.

STEFANI, S. et al. Prevalence and characterisation of extended-spectrum p- lactamase-producing Enterobacteriaceae in food-producing animals in Northern Italy. **New Microbiol,** v. 37, n. 4, p. 551-5, 2014.

TAMMA, P. D. et al. Carbapenem Therapy is Associated with Improved Survival Compared to Piperacillin-Tazobactam for Patients with ESBL Bacteremia. **Clinicai Infectious Diseases,** p. civ003, 2015.

TORTORA, G. J.; FUNKE, B. R.; CASE, C. L. **Microbiology.** 10. ed. Porto Alegre: Artmed, p. 934, 2010.

UMADEVI, S. et al. Prevalence and antimicrobial susceptibility pattern of ESBL producing Gram

Negative Bacilli. **Journal of Clinicai and Diagnostic Research,** v. 5, n. 2, p. 236-239, 2011.

WANG, G. et al. CTX-M P-Lactamase-producing Klebsiella pneumoniae in Suburban New York City, New York, USA. **Emerging infectious diseases,** v. 19, n. 11, p. 1803-1810, 2013.

WELLINGTON, E. M. H. et al. The role of the natural environment in the emergence of antibiotic resistance in Gram-negative bacteria. **The Lancet infectious diseases,** v. 13, n. 2, p. 155-165, 2013.

WILLIAMS, J. D. β-lactamases and β-lactamase inhibitors. Inter. J. Antimicrob. Agents, **Elsevier Science B.V. University of London.** London n.12. p. 3-7, Aug.1999.

ZARFEL, G. et al. Comparison of extended-spectrum-p-lactamase (ESBL) carrying Escherichia coli from sewage sludge and human urinary tract infection. **Environmental Pollution,** v. 173, p. 192-199, feb. 2013.

CHAPTER 7

APPENDIX

Figure 17- Species carrying the CTX-M gene and resistance profile to the antimicrobials tested in the antibiogram using the disc diffusion technique

N°	SPECIES	AMI	ASB	AMP	ATM	CEFAZO	CAZ	CRO	CEFOT	CEFOX	CEFEP	IMP	LIGHT	PPT	TETRA	SUT	GEN	CLO	CFL	SOMA ATB
1	***Klebsiella ozaenae***	1	0	0	0	0	0	0	0	0	0	1	0	1	1	0	1	0	0	5
2	***Klebsiella ozaenae***	1	0	0	0	1	0	0	0	0	0	1	0	1	1	0	0	1	0	6
3	***Klebsiella ozaenae***	1	0	0	0	0	0	0	0	0	0	1	1	1	0	0	0	0	0	4
4	***Klebsiella ozaenae***	1	0	0	0	0	0	0	0	0	0	1	1	1	0	0	0	0	0	4
5	***Klebsiella ozaenae***	1	0	0	1	1	0	1	1	0	0	1	1	1	0	0	0	1	0	9
6	***Klebsiella ozaenae***	0	0	0	0	0	0	0	0	0	0	1	1	1	0	0	0	0	0	3
7	***Klebsiella sp***	0	0	0	0	0	0	0	0	0	0	0	0	1	0	0	0	0	0	1
8	***Klebsiella sp***	1	0	0	0	0	0	0	0	0	0	1	0	1	1	0	0	0	0	4
9	***Klebsiella sp***	1	0	0	0	0	0	0	0	0	0	1	0	1	0	0	0	1	0	4
10	***Klebsiella sp***	0	0	0	0	0	0	0	0	0	0	1	0	1	0	0	0	1	0	3
11	***Klebsiella sp***	0	0	0	0	0	0	0	0	0	0	1	0	1	0	0	0	0	0	2
12	***Klebsiella rhynoscleromatis***	1	0	0	0	0	0	0	0	0	0	1	0	1	0	0	0	1	0	4
13	***Klebsiella pneumoniae***	1	0	0	0	0	0	0	0	0	0	1	1	1	0	0	0	0	0	4
14	***Enterobacter sp***	0	0	0	0	0	0	0	0	0	0	1	1	1	0	0	0	1	0	4
15	***Enterobacter sp***	0	0	0	0	0	0	0	0	0	0	1	1	1	0	0	0	0	0	3
16	***Enterobacter cloacae***	0	0	0	0	0	0	0	0	0	0	1	1	1	0	0	1	1	0	5
17	**BGN**	0	0	0	0	0	0	0	0	0	0	0	0	0	0	0	0	0	0	0
18	**BGN**	0	0	0	0	0	0	0	0	0	0	0	0	0	0	0	0	0	0	0
19	***Escherichia coli***	0	0	0	0	0	0	0	0	0	0	1	1	1	1	0	0	1	0	5
20	***Escherichia coli***	0	0	0	0	0	0	0	0	0	0	1	0	1	1	0	0	1	0	4
21	***Pseudomonas sp***	0	0	0	0	0	0	0	0	0	0	1	1	1	0	0	1	0	0	4
	TOTAL	9	0	0	1	2	0	1	1	0	0	18	10	19	5	0	3	9	0	
*0-Resistant, 1-Susceptible																				

AMI: Amikacin; ASB: Ampicillin/Sulbactam; AMP: Ampicillin; ATM: Aztreonam; CEFAZO: Cefazolin; CAZ: Ceftadizime; CRO: Ceftriaxone; CEFOT: Cefotaxime; CEFOX: Cefoxitin; CEFEP: Cefepime; IMP: Imipenem; LEVO: Levfloxacin; PPT: Piperacillin/Tazobactam; TETRA: Tetracycline; SUT: Sulfatrimethropime; GEN: Gentamicin; CLO: Chloramphenicol; CFL: Cephalexin.

ANNEXES

Annex 1 - First ANVISA (2013) table for determining the Biochemical Standard for identifying the main Enterobacteriaceae of clinical importance

Bacteria/Evidence	Indol	Citrate	H.S	Urea	Fenilala	Lysine	Arginine	Ornithine	Motil.
Citrobacter freundii	33	78	78	44	0	0	67	0	89
Citrobacter diversus (koseri)	99	99	0	75	0	0	80	99	95
Citrobacter amalonaticus	100	95	5	85	0	0	85	95	95
Edwarsiella tarda	99	1	100	0	0	100	0	100	98
Enterobacter aerogenes	0	95	0	2	0	98	0	98	97
Enterobacter doacae	0	100	0	65	0	0	97	96	95
Enterobacter gergoviae	0	99	0	93	0	90	0	100	90
Enterobacter sakazakii	11	99	0	1	50	0	99	91	96
Escherichia coti	98	1	1	1	0	90	17	65	95
Inactive Escherichia coli	80	1	1	1	0	40	3	20	5
Shigella dysenteriae	45	0	0	0	0	0	2	0	0
Shigella flexneri	50	0	0	0	0	0	5	0	0
Shigella boydii	25	0	0	0	0	0	10	2	0
Shigella sonnei	0	0	0	0	0	0	2	98	0
Hafnia alvei	0	10	0	4	0	100	6	98	85
Klebsiella pneumoniae	0	98	0	95	0	98	0	0	0
Klebsiella oxytoca	99	95	0	90	1	99	0	0	0
Klebsiella ozaenae	0	30	0	10	0	40	6	0	0
Klebsiella rhinoscleromatis	0	0	0	0	0	0	0	0	0
Morganella morganii group	95	0	20	95	95	V	0	95	V
Pantoea agglomerans	20	50	0	20	20	0	0	0	85
Proteus mirabilis	2	65	98	98	98	0	0	99	95
Proteus vulgaris	98	15	95	95	99	0	0	0	95
Proteus penneri	0	0	30	100	99	0	0	0	85
Providencia rettgeri	99	95	0	90	98	0	0	0	94
Providencia stuartii	98	93	0	30	95	0	0	0	85
Providencia alcalifaciens	99	90	0	0	90	0	0	0	96
Salmonella spp.	1	95	95	1	0	98	70	97	95
Salmonell typhi	0	0	97	0	0	98	3	0	97
Salmonella cholerasuis	0	25	50	0	0	95	55	100	95
Salmonella paratyphi A	0	0	10	0	0	0	15	95	95
Salmonella gallinarum	0	0	100	0	0	90	10	95	0
Salmonella pullorum	0	0	90	0	0	100	10	95	0
Salmonella others	1	90	100	0	0	99	70	99	99
Serratia marcescens	1	98	0	15	0	99	0	99	97
Serratia marcescens bio 1	0	30	0	0	0	55	4	65	14
Serratia liquefaciens	1	90	0	3	0	95	0	95	95
Serratia rubidae	0	95	0	2	0	55	0	0	85
Yersinia enterocolitica	50	0	0	75	0	0	0	95	2

Annex 2 - Second ANVISA (2013) table for determining the Biochemical Standard for identifying the main Enterobacteriaceae of clinical importance.

Bacteria/Evidence	gelatine	Malonato	Glucose gas	lactose	sucrose esculin		DNAse
Citrobacter freundii	0	11	89	78	89	0	0
Citrobacter diversus (koseri)	0	95	98	50	40	1	0
Citrobacter amalonaticus	0	1	97	35	9	5	0
Edwarsiella tarda	0	0	100	0	0	0	0
Enterobacter aerogenes	0	95	100	95	100	98	0
Enterobacter cloacae	0	75	100	93	97	30	0
Enterobacter gergoviae	0	96	98	55	98	97	0
Enterobacter sakazakii	0	18	98	99	100	100	0
Escherichia coli	0	0	95	95	50	35	0
Inactive Escherichia coli	0	0	5	25	15	5	0
Shigella dysenteriae	0	0	0	0	0	0	0
Shigella flexneri	0	0	3	1	1	0	0
Shigella boydii	0	0	0	1	0	0	0
Shigella sonnei	0	0	0	2	1	0	0
Hafnia alvei	0	50	98	5	10	7	0
Klebsiella pneumoniae	0	93	97	98	99	99	0
Klebsiella oxytoca	0	95	97	100	100	100	0
Klebsiella ozaenae	0	3	50	30	20	80	0
Klebsiella rhinoscleromatis	0	95	0	0	75	60	0
Morganella morganii group	0	1	99	1	0	0	0
Pantoea agglomerans	2	65	20	40	75	60	0
Proteus mirabilis	90	2	90	2	15	0	50
Proteus vulgaris	91	0	85	2	97	50	80
Proteus penneri	50	0	45	1	100	0	40
Providencia rettgeri	0	0	10	5	15	35	0
Providencia stuartii	0	0	0	2	50	0	10
Providencia alcalifaciens	0	0	85	0	15	0	0
Salmonella spp.	0	0	90	1	1	5	2
Salmonell typhi	0	0	0	1	0	0	0
Salmonella cholerasuis	0	0	95	0	0	0	0
Salmonella paratyphi /I	0	0	99	0	0	0	0
Salmonella gallinarum	0	0	0	0	0	0	10
Salmonella pullorum	0	0	90	0	0	0	0
Salmonella others	1	V	100	V	1	Or 15	1
Serratia marcescens	90	3	55	2	99	95	98
Serratia marcescens bio 1	30	0	0	4	100	90	82
Serratia liquefaciens	90	2	75	10	98	97	85
Serratia rubidae	90	94	30	100	99	94	99
Yersinia enterocolitica	0	6	5	5	95	25	5

Source: ANVISA (2013).

Legend: V = variable

MIX
Papier aus verantwortungsvollen Quellen
Paper from responsible sources
FSC® C105338
FSC
www.fsc.org

Printed by Books on Demand GmbH, Norderstedt / Germany